The Ironstone Garden Journal

The All-in-One Recordkeeping System for Gardeners of Fruits, Flowers, Herbs & Vegetables

Property of:

____

____

AF874414

ISBN: 978-1-968971-14-4

Now the Lord God had planted a garden in the east, in Eden; and there he put the man he had formed. The Lord God made all kinds of trees grow out of the ground—trees that were pleasing to the eye and good for food.
Genesis 2:8-9

Ironstone Books

Disclaimer and Terms of Use:

The author and publisher have made every effort to ensure the completeness of this record book. However, they do not guarantee that all information is error-free or entirely comprehensive.

While diligent efforts have been made to verify the content, the author and publisher are not responsible for any errors, omissions, or differing interpretations of the material presented. Any unintended references that may appear to slight individuals, businesses, or organizations are purely coincidental.

This content may not be reproduced in any form without prior written permission from the author.

Copies of this book are available for book signings, special events, and retail gift shop purchases. For direct ship information, of copies of 10 or more, please contact the author at ironstone_tennessee@yahoo.com.

The Ironstone Garden Journal
The All-in-One Recordkeeping System for Gardeners of Fruits, Flowers, Herbs & Vegetables

Printed in the United States of America.

First printing, 2025

ISBN: 978-1-968971-14-4

Ironstone Tennessee

Calhoun Tennessee, 37309

www.ironstone-tennessee.com

Table of Contents

How to use The Ironstone Garden Journal

I want to thank you so much for choosing to purchase *The Ironstone Garden Journal*. I welcome feedback and suggestions for ways to make it more useful for future editions.

The Journal is designed to help you record your gardening journey. There is space to record the things you observe such as weather, bloom times, pollinators, pests and diseases, and the harvest times for your fruits and vegetables. There is also space to record the things that you do in the garden, such as planting, pruning, fertilizing, disease and pest treatments, mulching, harvesting, and major projects.

Important Dates and Data:

Record your **USDA growing zone, average and observed frost dates, average annual rainfall, soil type and pH here.** This information allows you to choose plants that will be most likely to do well in your climate. Most seed packets, plant labels and catalog listings will tell you the plant's range of growing zones.

Knowing your **average first and last frost dates** will help you to schedule appropriate shipping dates when purchasing live plants and trees online. It will also help you to schedule when to start seeds indoors, when it is (probably) safe to transplant outdoors, and when you will need to have raised beds and garden plots prepared for Spring and Fall planting.

Keeping in mind your **annual average precipitation** (snow and rain) will help you to choose what plants to grow based on your desired amount of time spent in the garden (and how big of a water bill you can afford). If you are a lazy gardener like me, or just don't have time in for your schedule for frequent watering, then choosing plants that can thrive with only rainfall (most of the time) makes sense.

Your **soil type and pH** will determine what amendments you may need, what gardening methods will be most successful for you, and what plants will and won't do well for you in the garden. A bit of quick Internet research will tell you whether a plant likes your type of soil and whether your soil pH is in the right range for it to thrive. Knowing this can save you a lot of money and work.

Goals

Before you start planning (or planting), it really helps to be clear and realistic about what your gardening goals are. Goals that are too big will doom you to frustration, discouragement and failure. They might just blow your budget too.

Why are you planting a garden? Are you doing a homeschool project with your kids? Maybe you want to get closer to nature? Are you trying to make your grocery budget go farther? Are you hoping to grow all the fruits and vegetables you eat? Learning a new skill? Looking for stress relief? Trying to reduce exposure to environmental toxins? Wanting to attract pollinators? Hoping to create a backyard oasis?

Whatever your big goal is, you need to keep it in mind as you work on your garden. Take an honest look at your available free time, tools, funds and skills. Some goals will cost you a lot more in time and money than others. Some goals are more forgiving of mistakes, changes of plans or delays than others.

Your smaller goals might be things like:
- Can salsa and spaghetti sauce for a year, made only from your own garden produce
- Grow herbs to create your own herbal tea blends
- Experiment with sweet potatoes
- Grow cherry tomatoes on your patio from seed

Before making any big decisions (and some of the small ones too) such as the purchase of a tool, the size of your garden, or how many tomato plants you need to buy, always ask yourself: "Does this help me to achieve my big goal?"

Working in Your Garden

Pruning Log

Record What trees and shrubs you prune, your reason for pruning (size, shape, increase blooms etc.) the date of pruning, and when you will need to prune again.

Seed Starting Record

It's easy to forget when you started your seeds as the days pass, so write it all down here so you don't forget. When you look back on your growing season, you'll be able to decide whether you need to start seeds earlier or later next time.

Fertilizer & Soil Amendments Record

Write down all of the details when you fertilize your plants or amend your soil. You'll be able to reference the data later on to determine whether your amendments were effective or not.

Pest & Disease Treatment Record

Note signs of pests and disease damage here, and how you decide to manage them. You'll be able to assess whether or not the treatment was effective, and note when a repeat treatment is needed.

Project Trackers

Whether you're building a gazebo, installing a pond, or planting a fruit orchard, you can keep track of materials, costs, helpers and progress here.

Permanent Beds

This space is most likely to be used for perennial flower beds, asparagus beds, foundation plantings or perennial herb beds. There is space to record what plants are growing in the beds, and your observations about them. The graph can be used to map the outline of the bed and diagram the position of each plant.

Container Garden

Here you can detail what you are growing in your containers. Whether it's hanging baskets of annuals, dwarf potted fruit trees, or an arrangement of potted herbs, you can take notes on your container plants here.

4x8 & 3x9 Raised Beds

This space is for documenting how your raised bed plantings are laid out, what plants are growing in each raised bed, and how you are managing them. If your raised beds are a different size, you can simply outline the part of the grid you need and disregard the rest.

Garden Rows & High Tunnel

This space allows you to draw out your row crops or those grown in a high tunnel. If space allows, you might even be able to design two different versions of each area for succession planting or future crop rotations. The lined pages have been purposely left unstructured so that you can organize them to suit your needs. They can be used for a 6x8 greenhouse or a 30x70 high tunnel, a 10x10 garden plot or a 2-acre garden.

Planting Record

This space allows you to document when you set out seedlings, direct sow seeds or plant bare root trees, plants or shrubs. It's all too easy to forget these dates, and I should know, I am an expert forgetter (it's kind of why I created this journal).

Local Seed & Plant Purchases

Here is a place to record the seeds and plants you purchase or trade from local garden centers, big box stores, marketplace sellers, friends and neighbors. You might remember what you paid or where you got something a year from now, but memory fades over time, and in five or ten years, you may have forgotten. If it's written down, you don't have to remember it.

Online and Catalog Seed & Plant Purchases

Some seeds and plants aren't available locally, so here is space to record your purchases from online and catalog retailers.

Book, Tool & Supplies Purchases

This is a separate space to record the non-plant items you purchase for your garden. It may seem silly, but I've repurchased a book that I forgot I already owned more than once, and I don't always remember where my tools and soil amendments came from. By writing it down, I'm able to share that information with other gardeners.

Harvest Record

You can use these charts in a couple of different ways. You can either mark an x in the space that corresponds with each week that a particular plant produces a harvest, or you can write down how much you harvest from each plant in a given week. You can use a page for each type of plant (one for apple trees, one for herbs, one for annual vegetables etc.) or for each 3-month period you can list each plant as you begin to harvest from it. If you aren't growing food plants, you can use this space to note when your flowers go to seed, perennials are divided, when the bloom season ends, or any other seasonal markers that you choose. You can even use these records to note harvest times of plants you are considering for future planting, so you can plan out your harvest season.

Observing Your Garden

Month-at-a-Glance Calendars

This space is for noting major events in your gardening year. You might note when the hummingbirds arrive in your area, when the Japanese beetles first appear, phases of the moon, serious weather events, frost damage, etc. It's also a good place to put reminders to yourself of things to be watching for, such as seed germination, when to reapply a fertilizer or pest control method, floating row covers etc. It's your journal, do what makes your life easier.

Weather Log

This space allows your to record daily high and low temperatures, precipitation, wind, and cloud cover. Tracking weather trends can help you to determine when you need to do supplemental watering or when you need to protect tender plants from frost or sun damage. It can also help you to determine why a crop is delayed or why insect pressure is especially heavy or light from one year to the next.

Pollinator Notes

This space is for you to observe the arrival and activity of pollinators in your garden. Their activity can be a huge benefit to your garden, and observing them can help you spot problems and assess the overall health of your garden.

Weed Notes

As much of a nuisance as they can be, weeds do serve a purpose in your garden. They cover bare soil and prevent erosion. Some are edible or medicinal, while others are toxic. The presence of certain weeds can signal problems with soil health or pH. Learning to pay attention to your weeds can help you to become a better gardener.

Bloom Times

Whether you're looking at your apple trees, your roses or your summer squash, taking note of the bloom time of your plants gives you some valuable information.

- Fruit trees whose blossoms are damaged by a late frost will produce little or no fruit that year. But that same fruit tree might also not produce fruit if it doesn't have another fruit tree to cross-pollinate. If your fruit tree isn't producing fruit, it's important to know why so you can take steps to remedy the problem.

- If all of the flowers in an ornamental bed bloom in June, with nothing the rest of the year, you might choose to add flowers that bloom at other times so the bed is attractive throughout the growing season.
- If you note that your tomato plants stop blooming the last week July due to intense heat, you won't be surprised by a corresponding lack of tomatoes for a week in the middle of your tomato harvest.
- You might even be able to determine what trees or plants are setting off your seasonal allergies.

Crop Yields

This space is to record how much food your garden produces. This information can help you plan next year's garden so that you're not drowning in more produce than you can handle, or disappointed because you didn't plant enough jalapenos to can a year's supply of salsa. You can record crop yields by type of crop (one page for tomatoes, one for root crops etc), or by harvest season.

Orchard Notes

This space is for making observations about your fruit and nut trees. You might notice that a robin has built a nest in your peach tree, or that there are signs of rabbit or deer damage to your apple trees. Maybe you see a couple of trees that need to be staked and pulled straight, or that the fruit on your pear trees needs thinning. These notes can be compared from one year to the next to remind you to schedule maintenance, chores or to be on the lookout for problems.

Journaling Your Garden

Daily Journal

This is the largest section, requiring the least amount of explanation. Each page has space for two days of journal entries. Depending on how often you work in your garden, there could be anywhere from 1-4 years worth of journal space. I would expect that unless you are a market gardener in a very temperate climate, you will get closer to 4 years of use from this journal. You can note what you did in the garden, what you noticed, what you need to do in the coming days, or even just how you feel about the time you spent in your garden. Again, this is your journal, use it however you need to.

It is my hope that you will find this journal to be the tool that helps you to grow a more successful and enjoyable garden from year to year.

Notes:

Important Dates and Data

Growing Zone	Soil Type	Soil pH	Avg. Last Frost	Avg. 1st Frost	Avg. Rainfall

My Gardening Challenges:

My Gardening Goals:

Notes:

Working in Your Garden

Pruning Log

Tree or Shrub	Date	Pruning Goal	Repeat Date	Notes:

Pruning Log

Tree or Shrub	Date	Pruning Goal	Repeat Date	Notes:

Seed Starting Record

Plant	Cultivar	#Seeds Sown	Sow Date	Germ. Date	# Germ.	Plant Out Date

Seed Starting Record

Plant	Cultivar	#Seeds Sown	Sow Date	Germ. Date	# Germ.	Plant Out Date

Seed Starting Record

Plant	Cultivar	#Seeds Sown	Sow Date	Germ. Date	# Germ.	Plant Out Date

Seed Starting Record

Plant	Cultivar	#Seeds Sown	Sow Date	Germ. Date	# Germ.	Plant Out Date

Fertilizer & Soil Amendment Record

Plant/Area	Date	Fertilizer	Repeat Date	Notes

Fertilizer & Soil Amendment Record

Plant/Area	Date	Fertilizer	Repeat Date	Notes

Pest and Disease Control Record

Date	Pest or Disease	Affected Plant(s)	Treatment	Repeat Date	Withdrawal Time	Effective?

Pest and Disease Control Record

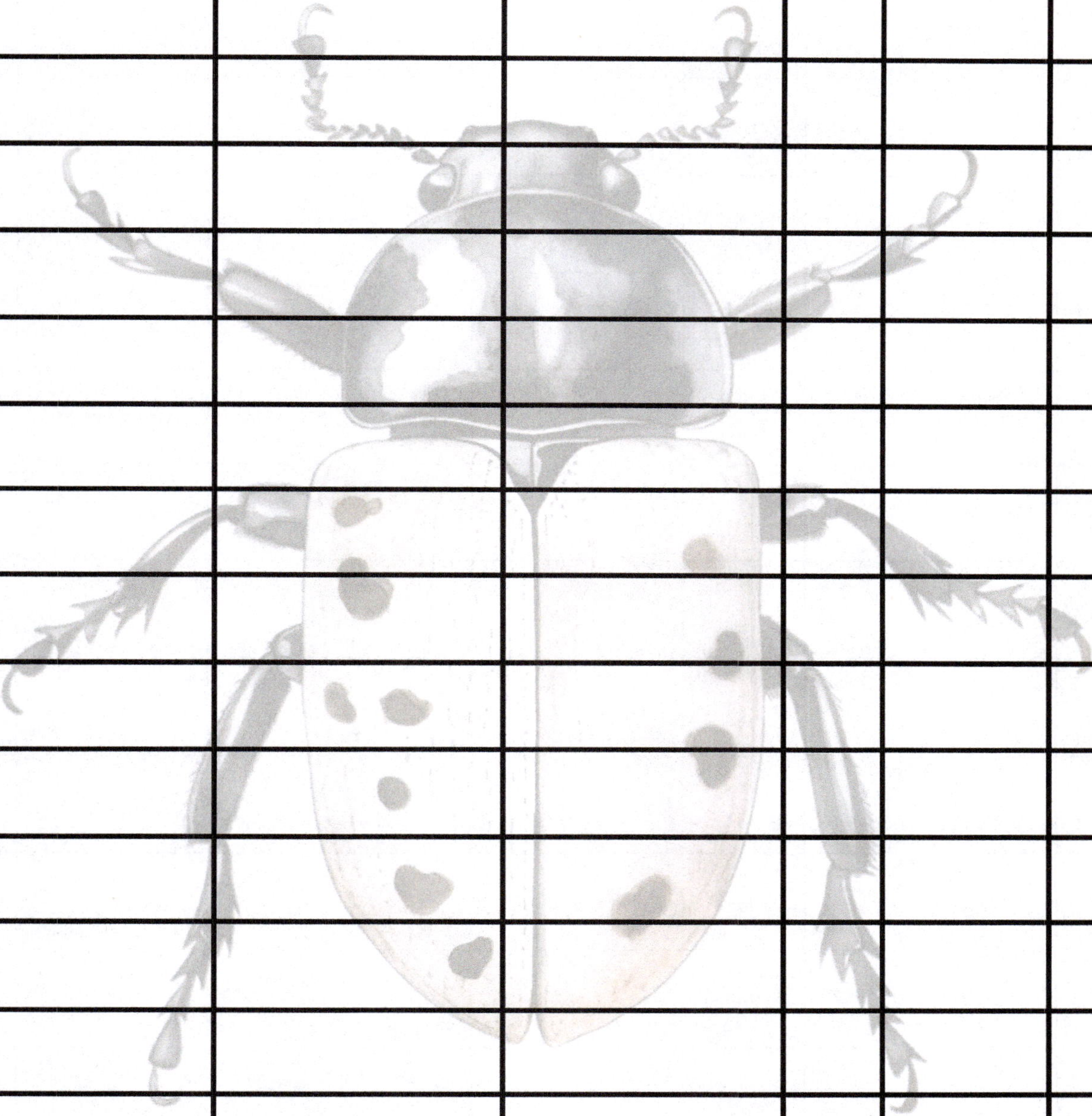

Date	Pest or Disease	Affected Plant(s)	Treatment	Repeat Date	Withdrawal Time	Effective?

Project Tracker:

Date	Materials	Time	Problems	Progress	Helper(s)	$ Spent

Project Tracker:

Date	Materials	Time	Problems	Progress	Helper(s)	$ Spent

Project Tracker:

Date	Materials	Time	Problems	Progress	Helper(s)	$ Spent

Project Tracker:

Date	Materials	Time	Problems	Progress	Helper(s)	$ Spent

Project Tracker:

Date	Materials	Time	Problems	Progress	Helper(s)	$ Spent

Project Tracker:

Date	Materials	Time	Problems	Progress	Helper(s)	$ Spent

Project Tracker:

Date	Materials	Time	Problems	Progress	Helper(s)	$ Spent

Project Tracker:

Date	Materials	Time	Problems	Progress	Helper(s)	$ Spent

Permanent Bed: ______________________________

Plants		Notes

Permanent Bed: ______________________________

Plants		Notes

Permanent Bed: ______________________________

Plants		Notes

Permanent Bed: ______________________________

Plants		Notes

Permanent Bed: ______________________________

Plants		Notes

Permanent Bed: ____________________

Plants		Notes

Permanent Bed: ____________________

Plants		Notes

Permanent Bed: ____________________

Plants		Notes

Container Garden

Container:________________ Date:__/___

Water:	Plants:
Observations:	
Harvest:	

Container:________________ Date:__/___

Water:	Plants:
Observations:	
Harvest:	

Container:________________ Date:__/___

Water:	Plants:
Observations:	
Harvest:	

Container:________________ Date:__/___

Water:	Plants:
Observations:	
Harvest:	

Container Garden

Container:________________ Date:__/___

Water:	Plants:
Observations:	
Harvest:	

Container:________________ Date:__/___

Water:	Plants:
Observations:	
Harvest:	

Container:________________ Date:__/___

Water:	Plants:
Observations:	
Harvest:	

Container:________________ Date:__/___

Water:	Plants:
Observations:	
Harvest:	

Container Garden

Container:_______________ Date:__/___

Water:	Plants:
Observations:	
Harvest:	

Container:_______________ Date:__/___

Water:	Plants:
Observations:	
Harvest:	

Container:_______________ Date:__/___

Water:	Plants:
Observations:	
Harvest:	

Container:_______________ Date:__/___

Water:	Plants:
Observations:	
Harvest:	

Container Garden

Container:________________ Date:__/___

Water:	Plants:
Observations:	
Harvest:	

Container:________________ Date:__/___

Water:	Plants:
Observations:	
Harvest:	

Container:________________ Date:__/___

Water:	Plants:
Observations:	
Harvest:	

Container:________________ Date:__/___

Water:	Plants:
Observations:	
Harvest:	

Container Garden

Container:________________ Date:__/___

Water:	Plants:
Observations:	
Harvest:	

Container:________________ Date:__/___

Water:	Plants:
Observations:	
Harvest:	

Container:________________ Date:__/___

Water:	Plants:
Observations:	
Harvest:	

Container:________________ Date:__/___

Water:	Plants:
Observations:	
Harvest:	

Container Garden

Container:________________ Date:__/___

Water:	Plants:
Observations:	
Harvest:	

Container:________________ Date:__/___

Water:	Plants:
Observations:	
Harvest:	

Container:________________ Date:__/___

Water:	Plants:
Observations:	
Harvest:	

Container:________________ Date:__/___

Water:	Plants:
Observations:	
Harvest:	

4 x 8 *Raised Beds*

Bed ______________________________

		Plants
Soil		
Mulch		
Watering		
Fertilizer		
Plant Date		
Problems		

Bed ______________________________

		Plants
Soil		
Mulch		
Watering		
Fertilizer		
Plant Date		
Problems		

4 *x* 8 ***Raised Beds***

Bed________________________________

Bed________________________________

4 x 8 *Raised Beds*

Bed ______________________________

Soil	
Mulch	
Watering	
Fertilizer	
Plant Date	
Problems	

Plants

Bed ______________________________

Soil	
Mulch	
Watering	
Fertilizer	
Plant Date	
Problems	

Plants

4 x 8 *Raised Beds*

Bed________________________________

Bed________________________________

4 x 8 *Raised Beds*

Bed ________________________________

		Plants
Soil		
Mulch		
Watering		
Fertilizer		
Plant Date		
Problems		

Bed ________________________________

		Plants
Soil		
Mulch		
Watering		
Fertilizer		
Plant Date		
Problems		

4 x 8 *Raised Beds*

Bed______________________________

Bed______________________________

4 x 8 *Raised Beds*

Bed ______________________________

Soil	
Mulch	
Watering	
Fertilizer	
Plant Date	
Problems	

Plants

Bed ______________________________

Soil	
Mulch	
Watering	
Fertilizer	
Plant Date	
Problems	

Plants

4 x 8 *Raised Beds*

Bed ______________________________

Bed ______________________________

4 x 8 *Raised Beds*

Bed ______________________________

		Plants
Soil		
Mulch		
Watering		
Fertilizer		
Plant Date		
Problems		

Bed ______________________________

		Plants
Soil		
Mulch		
Watering		
Fertilizer		
Plant Date		
Problems		

4 x 8 *Raised Beds*

Bed ______________________________

Bed ______________________________

3 x 9 *Raised Beds*

Bed ____________________________

		Plants
Soil		
Mulch		
Watering		
Fertilizer		
Plant Date		
Problems		

Bed ____________________________

		Plants
Soil		
Mulch		
Watering		
Fertilizer		
Plant Date		
Problems		

Bed ____________________________

		Plants
Soil		
Mulch		
Watering		
Fertilizer		
Plant Date		
Problems		

3 x 9 *Raised Beds*

Bed ______________________________

Bed ______________________________

Bed ______________________________

3 x 9 *Raised Beds*

Bed______________________________

Soil		Plants
Mulch		
Watering		
Fertilizer		
Plant Date		
Problems		

Bed______________________________

Soil		Plants
Mulch		
Watering		
Fertilizer		
Plant Date		
Problems		

Bed______________________________

Soil		Plants
Mulch		
Watering		
Fertilizer		
Plant Date		
Problems		

3 x 9 *Raised Beds*

Bed______________________________

Bed______________________________

Bed______________________________

3 x 9 *Raised Beds*

Bed ______________________________

Soil		Plants
Mulch		
Watering		
Fertilizer		
Plant Date		
Problems		

Bed ______________________________

Soil		Plants
Mulch		
Watering		
Fertilizer		
Plant Date		
Problems		

Bed ______________________________

Soil		Plants
Mulch		
Watering		
Fertilizer		
Plant Date		
Problems		

3 x 9 *Raised Beds*

Bed________________________________

Bed________________________________

Bed________________________________

3 x 9 *Raised Beds*

Bed ______________________________

Soil		Plants
Mulch		
Watering		
Fertilizer		
Plant Date		
Problems		

Bed ______________________________

Soil		Plants
Mulch		
Watering		
Fertilizer		
Plant Date		
Problems		

Bed ______________________________

Soil		Plants
Mulch		
Watering		
Fertilizer		
Plant Date		
Problems		

3 x 9 *Raised Beds*

Bed________________________________

Bed________________________________

Bed________________________________

Garden Rows or High Tunnel

Garden Rows or High Tunnel

Garden Rows or High Tunnel

Garden Rows or High Tunnel

Garden Rows or High Tunnel

Garden Rows or High Tunnel

Planting Record

	Month:____________				Month:____________				Month:____________			
Plant	1	2	3	4	1	2	3	4	1	2	3	4

Planting Record

	Month:____________				Month:____________				Month:____________			
Plant	1	2	3	4	1	2	3	4	1	2	3	4

Planting Record

	Month:__________				Month:__________				Month:__________			
Plant	1	2	3	4	1	2	3	4	1	2	3	4

Planting Record

	Month:____________				Month:____________				Month:____________			
Plant	1	2	3	4	1	2	3	4	1	2	3	4

Planting Record

	Month:__________				Month:__________				Month:__________			
Plant	1	2	3	4	1	2	3	4	1	2	3	4

Planting Record

	Month:________				Month:________				Month:________			
Plant	1	2	3	4	1	2	3	4	1	2	3	4

Planting Record

	Month:____________				Month:____________				Month:____________			
Plant	1	2	3	4	1	2	3	4	1	2	3	4

Planting Record

	Month:____________				Month:____________				Month:____________			
Plant	1	2	3	4	1	2	3	4	1	2	3	4

Local Seed & Plant Purchases

Plant	Source	Purchase Date	Cost	Plant Date	Location

Local Seed & Plant Purchases

Plant	Source	Purchase Date	Cost	Plant Date	Location

Online/Catalog Seed & Plant Purchases

Plant	Source	Purchase Date	Ship Date	Cost	Plant Date	Location

Online/Catalog Seed & Plant Purchases

Plant	Source	Purchase Date	Ship Date	Cost	Plant Date	Location

Book, Tool & Supply Purchases

Book/Product/Tool	Source	Purpose/Subject	Date	Cost

Book, Tool & Supply Purchases

Book/Product/Tool	Source	Purpose/Subject	Date	Cost

Harvest Record

	Month:____________				Month:____________				Month:____________			
Plant	1	2	3	4	1	2	3	4	1	2	3	4

Harvest Record

	Month:____________				Month:____________				Month:____________			
Plant	1	2	3	4	1	2	3	4	1	2	3	4

Harvest Record

	Month:____________				Month:____________				Month:____________			
Plant	1	2	3	4	1	2	3	4	1	2	3	4

Harvest Record

	Month:____________				Month:____________				Month:____________			
Plant	1	2	3	4	1	2	3	4	1	2	3	4

Harvest Record

	Month:__________				Month:__________				Month:__________			
Plant	1	2	3	4	1	2	3	4	1	2	3	4

Harvest Record

	Month:__________				Month:__________				Month:__________			
Plant	1	2	3	4	1	2	3	4	1	2	3	4

Harvest Record

	Month:___________				Month:___________				Month:___________			
Plant	1	2	3	4	1	2	3	4	1	2	3	4

Harvest Record

	Month:__________				Month:__________				Month:__________			
Plant	1	2	3	4	1	2	3	4	1	2	3	4

Month: ____________________

Sunday	Monday	Tuesday	Wednesday

Notes:

Thursday	Friday	Saturday

Month: ____________

Sunday	Monday	Tuesday	Wednesday

Notes:

Thursday	Friday	Saturday

Month: ______________________

Sunday	Monday	Tuesday	Wednesday

Thursday	Friday	Saturday

Notes:

Month: ____________

Sunday	Monday	Tuesday	Wednesday

Notes:

Thursday	Friday	Saturday

Month: ____________

Sunday	Monday	Tuesday	Wednesday

Notes:

Thursday	Friday	Saturday

Month: ______________________

Sunday	Monday	Tuesday	Wednesday

Notes:

Thursday	Friday	Saturday

Month: ______________________

Sunday	Monday	Tuesday	Wednesday

Notes:

Thursday	Friday	Saturday

Month: ______________________

Sunday	Monday	Tuesday	Wednesday

Notes:

Thursday	Friday	Saturday

Month: ______________________

Sunday	Monday	Tuesday	Wednesday

Notes:

Thursday	Friday	Saturday

Month: ____________________

Sunday	Monday	Tuesday	Wednesday

Notes:

Thursday	Friday	Saturday

Month: ______________________________

Sunday	Monday	Tuesday	Wednesday

Notes:

Thursday	Friday	Saturday

Month: ____________________

Sunday	Monday	Tuesday	Wednesday

Notes:

Thursday	Friday	Saturday

Notes:

Observing Your Garden

Weather Log

High & Low Temperatures | Precipitation | Wind Speed & Direction | Cloud Cover

Date					Date				

Weather Log

High & Low Temperatures — Precipitation — Wind Speed & Direction — Cloud Cover

Date					Date				

Weather Log

High & Low Temperatures — Precipitation — Wind Speed & Direction — Cloud Cover

Date					Date				

Weather Log

High & Low Temperatures | Precipitation | Wind Speed & Direction | Cloud Cover

Date					Date				

Weather Log

High & Low Temperatures — Precipitation — Wind Speed & Direction — Cloud Cover

Date					Date				

Weather Log

High & Low Temperatures — Precipitation — Wind Speed & Direction — Cloud Cover

Date					Date				

Weather Log

High & Low Temperatures

Precipitation

Wind Speed & Direction

Cloud Cover

Date					Date				

Weather Log

High & Low Temperatures   Precipitation   Wind Speed & Direction   Cloud Cover

Date					Date				

Weather Log

High & Low Temperatures

Precipitation

Wind Speed & Direction

Cloud Cover

Date					Date				

Weather Log

High & Low Temperatures | Precipitation | Wind Speed & Direction | Cloud Cover

Date					Date				

Weather Log

High & Low Temperatures | Precipitation | Wind Speed & Direction | Cloud Cover

Date	High & Low Temperatures	Precipitation	Wind Speed & Direction	Cloud Cover	Date	High & Low Temperatures	Precipitation	Wind Speed & Direction	Cloud Cover

Weather Log

High & Low Temperatures — Precipitation — Wind Speed & Direction — Cloud Cover

Date					Date				

Weather Log

High & Low Temperatures — Precipitation — Wind Speed & Direction — Cloud Cover

Date	High & Low Temperatures	Precipitation	Wind Speed & Direction	Cloud Cover	Date	High & Low Temperatures	Precipitation	Wind Speed & Direction	Cloud Cover

Weather Log

High & Low Temperatures | Precipitation | Wind Speed & Direction | Cloud Cover

Date					Date				

Weather Log

High & Low Temperatures   Precipitation   Wind Speed & Direction   Cloud Cover

Date					Date				

Weather Log

High & Low Temperatures
Precipitation
Wind Speed & Direction
Cloud Cover

Date					Date				

Pollinator Notes

Pollinator	Date	Plant

Pollinator Notes

Pollinator	Date	Plant

Weed Notes

Weed	Location	Date	Signals/Uses/Danger

Weed Notes

Weed	Location	Date	Signals/Uses/Danger

Bloom Time Record

	Month:__________				Month:__________				Month:__________			
Plant	1	2	3	4	1	2	3	4	1	2	3	4

Bloom Time Record

	Month:____________				Month:____________				Month:____________			
Plant	1	2	3	4	1	2	3	4	1	2	3	4

Bloom Time Record

	Month:__________				Month:__________				Month:__________			
Plant	1	2	3	4	1	2	3	4	1	2	3	4

Bloom Time Record

	Month:____________				Month:____________				Month:____________			
Plant	1	2	3	4	1	2	3	4	1	2	3	4

Bloom Time Record

	Month:____________				Month:____________				Month:____________			
Plant	1	2	3	4	1	2	3	4	1	2	3	4

Bloom Time Record

	Month:__________				Month:__________				Month:__________			
Plant	1	2	3	4	1	2	3	4	1	2	3	4

Bloom Time Record

	Month:____________				Month:____________				Month:____________			
Plant	1	2	3	4	1	2	3	4	1	2	3	4

Bloom Time Record

	Month:____________				Month:____________				Month:____________			
Plant	1	2	3	4	1	2	3	4	1	2	3	4

Crop Yields

Crop	1st	2nd	3rd	4th	5th	6th

Crop Yields

Crop	1st	2nd	3rd	4th	5th	6th

Crop Yields

Crop	1st	2nd	3rd	4th	5th	6th

Crop Yields

Crop	1st	2nd	3rd	4th	5th	6th

Crop Yields

Crop	1st	2nd	3rd	4th	5th	6th

Crop Yields

Crop	1st	2nd	3rd	4th	5th	6th

Crop Yields

Crop	1st	2nd	3rd	4th	5th	6th

Crop Yields

Crop	1st	2nd	3rd	4th	5th	6th

Crop Yields

Crop	1st	2nd	3rd	4th	5th	6th

Crop Yields

Crop	1st	2nd	3rd	4th	5th	6th

Crop Yields

Crop	1st	2nd	3rd	4th	5th	6th

Crop Yields

Crop	1st	2nd	3rd	4th	5th	6th

Orchard Notes

Date:	Observations

Orchard Notes

Date:	Observations

Orchard Notes

Date:	Observations

Orchard Notes

Date:	Observations

Orchard Notes

Date:	Observations

Orchard Notes

Date:	Observations

Orchard Notes

Date:	Observations

Orchard Notes

Date:	Observations

Notes:

Journaling Your Garden

Date: / /

Date: / /

Date: / /

Date: / /

Date: / /

Date: / /

Date: / /

Date: / /

Date: / /

Date: / /

Date: / /

Date: / /

Date: / /

Date: / /

Date: / /

Date: / /

Date: / /

Date: / /

Date: / /

Date: / /

Date: / /

Date: / /

Date: / /

Date: / /

Date: / /

Date: / /

Date: / /

Date: / /

Date: / /

Date: / /

Date: / /

Date: / /

Date: / /

Date: / /

Date: / /

Date: / /

Date: / /

Date: / /

Date: / /

Date: / /

Date: / /

Date: / /

Date: / /

Date: / /

Date: / /

Date: / /

Date: / /

Date: / /

Date: / /

Date: / /

Date: / /

Date: / /

Date: / /

Date: / /

Date: / /

Date: / /

Date: / /

Date: / /

Date: / /

Date: / /

Date: / /

Date: / /

Date: / /

Date: / /

Date: / /

Date: / /

Date: / /

Date: / /

Date: / /

Date: / /

Date: / /

Date: / /

Date: / /

Date: / /

Date: / /

Date: / /

Date: / /

Date: / /

Date: / /

Date: / /

Date: / /

Date: / /

Date: / /

Date: / /

Date: / /

Date: / /

Date: / /

Date: / /

Date: / /

Date: / /

Date: / /

Date: / /

Date: / /

Date: / /

Date: / /

Date: / /

Date: / /

Date: / /

Date: / /

Date: / /

Date: / /

Date: / /

Date: / /

Date: / /

Date: / /

Date: / /

Date: / /

Date: / /

Date: / /

Date: / /

Date: / /

Date: / /

Date: / /

Date: / /

Date: / /

Date: / /

Date: / /

Date: / /

Date: / /

Date: / /

Date: / /

Date: / /

Date: / /

Date: / /

Date: / /

Date: / /

Date: / /

Date: / /

Date: / /

Date: / /

Date: / /

Date: / /

Date: / /

Date: / /

Date: / /

Date: / /

Date: / /

Date: / /

Date: / /

Date: / /

Date: / /

Date: / /

Date: / /

Date: / /

Date: / /

Date: / /

Date: / /

Date: / /

Date: / /

Date: / /

Date: / /

Date: / /

Date: / /

Date: / /

Date: / /

Date: / /

Date: / /

Date: / /

Date: / /

Date: / /

Date: / /

Date: / /

Date: / /

Date: / /

Date: / /

Date: / /

Date: / /

Date: / /

Date: / /

Date: / /

Date: / /

Date: / /

Date: / /

Date: / /

Date: / /

Date: / /

Date: / /

Date: / /

Date: / /

Date: / /

Date: / /

Date: / /

Date: / /

Date: / /

Date: / /

Date: / /

Date: / /

Date: / /

Date: / /

Date: / /

Date: / /

Date: / /

Date: / /

Date: / /

Date: / /

Date: / /

Date: / /

Date: / /

Date: / /

Date: / /

Date: / /

Date: / /

Date: / /

Date: / /

Date: / /

Date: / /

Date: / /

Date: / /

Date: / /

Date: / /

Date: / /

Date: / /

Date: / /

Date: / /

Date: / /

Date: / /

Date: / /

Date: / /

Date: / /

Date: / /

Date: / /

Date: / /

Date: / /

Date: / /

Date: / /

Date: / /

Date: / /

Date: / /

Date: / /

Date: / /

Date: / /

Date: / /

Date: / /

Date: / /

Date: / /

Date: / /

Date: / /

Date: / /

Date: / /

Date: / /

Date: / /

Date: / /

Date: / /

Date: / /

Date: / /

Date: / /

Date: / /

Date: / /

Date: / /

Date: / /

Date: / /

Date: / /

Date: / /

Date: / /

Date: / /

Date: / /

Date: / /

Date: / /

Date: / /

Date: / /

Date: / /

Date: / /

Date: / /

Date: / /

Date: / /

Date: / /

Date: / /

Date: / /

Date: / /

Date: / /

Date: / /

Date: / /

Date: / /

Date: / /

Date: / /

Date: / /

Date: / /

Date: / /

Date: / /

Date: / /

Date: / /

Date: / /

Date: / /

Date: / /

Date: / /

Date: / /

Date: / /

Date: / /

Date: / /

Date: / /

Date: / /

Date: / /

Date: / /

Date: / /

Date: / /

Date: / /

Date: / /

Date: / /

Date: / /

Date: / /

Date: / /

Date: / /

Date: / /

Date: / /

Resources:

USDA Hardiness Zones: https://planthardiness.ars.usda.gov/

Mother Earth News: **https://www.motherearthnews.com/organic-gardening/**

Johhny's Select Seeds: **https://www.johnnyseeds.com/growers-library/methods- tools-supplies/growers-library-methods-tools-supplies.html**

Farmers' Almanac: **https://www.almanac.com/gardening**

UT Extension Master Gardener: **https://mastergardener.tennessee.edu/**

ATTRA: **https://attra.ncat.org/publication-library/**

University of Missouri Extension: **https://extension.missouri.edu/**

Rodale Institute: **https://rodaleinstitute.org/education/resources/**

Ironstone Tennessee Gardening Resources Page: **https://www.ironstone-tennessee.com/garden/gardening-resources**

Notes:

Melanie Holsti is an avid tea lover with a deep appreciation for the simple joys of life. Whether she's sipping a cup of Scottish Breakfast tea, capturing the perfect shot with her camera, or getting her hands dirty in the garden, she embraces each moment with enthusiasm.

As an author and small business owner, she pours her focus and creativity into everything she does.

Her background is as varied as her interests—she's worked in hospitality, retail, livestock farming, writing, and, of course, gardening. Over the years, she has experienced the triumphs and challenges that come with growing food and homesteading in diverse climates, from the harsh prairie of South Dakota to the woods and mountains of Missouri and Tennessee.

In 2018, Melanie and her family took on an exciting project—restoring a charming 1917 Craftsman bungalow. With a lot of hard work and a shared vision, they transformed it into the Ironstone Cottage Tea Room, where guests could gather and enjoy a hot cup of tea or lunch with friends in a cozy, welcoming atmosphere.

These days, Melanie, her husband, and their daughters reside on a small farm in Tennessee, living in a beautiful Federal-style home they built together. Their homestead is shared with an assortment of animals, including a peculiar flock of barn cats, free-ranging chickens, and noisy ducks.

When she's not tending to the garden or caring for her animals, Melanie finds joy in writing books, sampling new and exotic teas, and cheering on her favorite professional hockey team. No matter the season, she's always pursuing her interests and creating a life filled with warmth and, of course, plenty of tea.

www.ironstone_tennessee.com

IronstoneTennessee

IronstoneTennessee

amz

amazon.com/author/ironstonetennessee

ironstonetennessee

secret resources web page
Password: IronstoneBooks

Your feedback and reviews are priceless. If you have suggestions for improving my garden journals, I'd love to hear them. If you have a moment, please consider leaving a review on Amazon.

Other Titles by Melanie Holsti

The Ironstone Garden Planner

The Fruit, Flower, Herb & Vegetable Gardener's Planner

The Fruit, Flower, Herb & Vegetable Gardener's Journal

The Ornamental Files

The Houseplant Files

The Herb Files

The Orchard Files

The Vegetable Files

The Ironstone Forager's Journal

Journaling Your Weather

Ironstone Prayer Journals

Journaling Your Health (A collection of women's wellness journals)

Tea & Tisane Blending Workbook: A Tea Mixologist's Journal (with Jennifer C. Petersen)

Printed in Dunstable, United Kingdom

91335484R00185